JULY 2022

GENDER, TENURE SECURITY, AND LANDSCAPE GOVERNANCE

SYNTHESIS OF STUDIES FROM PIM'S GOVERNANCE OF NATURAL RESOURCES FLAGSHIP PROGRAM, 2013–2020

PATRICIA KRISTJANSON

A Peer-Reviewed Publication
International Food Policy Research Institute
Washington, DC

ABOUT IFPRI

The International Food Policy Research Institute (IFPRI), a CGIAR center established in 1975, provides research-based policy solutions to sustainably reduce poverty and end hunger and malnutrition. IFPRI's strategic research aims to foster a climate-resilient and sustainable food supply; promote healthy diets and nutrition for all; build inclusive and efficient markets, trade systems, and food industries; transform agricultural and rural economies; and strengthen institutions and governance. Gender is integrated in all the Institute's work. Partnerships, communications, capacity strengthening, and data and knowledge management are essential components to translate IFPRI's research from action to impact. The Institute's regional and country programs play a critical role in responding to demand for food policy research and in delivering holistic support for country-led development. IFPRI collaborates with partners around the world.

Photo credits: (cover) M. Farouk/Shutterstock

ISBN: 978-0-89629-434-9

DOI: https://doi.org/10.2499/9780896294349

ACKNOWLEDGMENT

This work was undertaken as part of, and funded by, the CGIAR Research Program on Policies, Institutions, and Markets (PIM) led by the International Food Policy Research Institute (IFPRI). PIM was in turn supported by these donors: http://www.pim.cgiar.org/donors/.

Gender relations shape women's and men's identities, norms, rules, and responsibilities. They influence people's access to, use, and management of land and other natural resources (Agarwal 1997; Fortmann and Rocheleau 1989; Feldstein and Poats 1990; Joekes et al. 1995). They also influence, and are influenced by, ownership, tenure, and user rights to land and forests (Meinzen-Dick et al. 1997; Colfer et al. 2016). Policies and interventions aimed at improving the governance of natural resources and landscapes need to benefit women as well as men. A failure to address differing needs, roles, and responsibilities of women and men, and underlying gender relations and how they are changing, reduces their effectiveness and risks perpetuating gender inequality (Agarwal 1994; Siqueira et al. 2021; FAO 2011).

A substantial body of research on these important issues comes from a research theme of the CGIAR Research Program on Policies, Institutions, and Markets (PIM) that focused on the governance of natural resources (Flagship 5, or PIM5). Research in Flagship 5 examined the policy and institutional foundations for improved natural resource management. It analyzed how tenure security affects sustainable management and how individuals, groups, and communities govern land, water, fish stocks, and forests. An important focus of this work is on *understanding who has what rights with respect to these resources, particularly women and members of marginalized groups; what their roles are in managing natural resources; and the livelihood benefits they receive*. Without a contextualized understanding of these questions, policies and practices can inadvertently exclude women, reinforce historic practices of gender injustice, or introduce new inequalities that worsen natural resource management and poverty. Thus, research teams in Flagship 5 worked with partners in governments, universities, the private sector, and civil society with the goal of identifying actions that can strengthen tenure rights of poor and marginalized people, particularly women, in order to improve governance of natural resources; and that can strengthen institutional arrangements for sustainable use of natural resources within shared landscapes.

This paper provides a synthesis of PIM5's contributions to knowledge about gender and its influence on tenure security and landscape governance. The focus of Flagship 5's research was on gender and social inclusion in relation to tenure (land and natural resources) and to landscape governance (for example, land use planning). This review focuses on a subset of the research studies undertaken in this program from 2013 to 2020. It discusses the scientific contributions to the broader wealth of related literature by analyzing and summarizing key lessons about gender from these studies, where possible, with respect to outcomes and impacts on natural resource management, food security, and poverty alleviation. This analysis will help researchers better communicate the results of the Flagship's research in order to inform policymakers and donors, and to influence strategies and interventions aimed at addressing challenges related to these three arenas.

APPROACH

While some research focuses on gender and tenure (see Quisumbing and Kumar 2014; Colfer et al. 2016; Giovarelli and Scalise 2019; Monterroso et al. 2019; Doss and Meinzen-Dick 2020), and some on gender and governance (see De Graaf et al. 2017; Sijapati Basnett et al. 2017; Buck et al. 2019; FAO 2019; McLain et al. 2019), the literature does not offer a single framework that explicitly links gender, tenure (security and benefits), and governance (whose ideas and objectives are included in rules and laws).

However, Doss and Meinzen-Dick's (2018) women and land tenure security framework provides a useful starting point for this synthesis. This framework is aimed at understanding factors that influence women's land tenure security. It consists of four components: 1) the context in which women's land rights exist, 2) threats and opportunities (that is, the factors that strengthen or weaken those rights); 3) the action arena, which includes the actors who can affect women's tenure security and the resources actors can draw on to achieve their desired outcomes; and 4) the desired outcome, which is women's land tenure security. Governance comes into play, as this framework can be used by decision-makers, for example, to develop new rules, regulation, policies, or legal reforms (that is, a change in governance) to strengthen women's rights to land. Ensuring women's involvement/gender equity in governance aspects shapes what the benefits look like and who benefits from the reforms.

Using this framework as a starting point for reviewing the body of knowledge related to the intersection of gender, tenure, and governance, we see three broad (and sometimes overlapping) questions emerge:

1. In terms of context: **What have we learned about how governance changes related to land tenure rules and regulations affect outcomes for women and contribute to gender equality?** Along with an improved understanding of key factors found within the overlapping issues of gender, governance, and tenure in different circumstances, this line of inquiry leads to evidence and recommendations for tenure policies and institutional arrangements that can enhance gender equality.

2. In terms of understanding the factors that strengthen or weaken rights: **What do gender analyses aimed at informing tenure or governance reforms show?** This includes, for example, analyses that look at what factors drive women's and men's tenure (in)security and those that have found there are different sources of tenure insecurity, and thus different solutions for strengthening it, for men and women. This approach underscores the importance of understanding gender norms and local contexts to inform land and natural resource tenure-focused interventions.

3. With respect to the actors and resources available: **What approaches, methods, frameworks, and data are available for addressing the gender, landscape governance, and land and natural resource tenure nexus of issues?** Reviews of existing knowledge and the development of frameworks and tools for comparative analysis of outcomes and impacts of interventions aimed at addressing the multiple challenges faced in this nexus are critical for evidence-based decision-making.

This paper synthesizes the evidence generated to help address these three questions through a wide range of studies and countries, focusing particularly on cases where we now have evidence about outcomes and impacts in the nexus of gender, tenure, and governance.

The synthesis covers the period from 2013 to mid-2020. The list of papers reviewed was compiled from the deliverables listed in Flagship 5's project reports (2013–2020) that flagged tenure and/or gender as key topics/research focus areas (205 products). The review included journal articles, working papers, reports, and briefs and, in a few cases, conference/workshop presentations or webinars and blogs. By focusing primarily on the articles, papers, and reports that addressed the three questions above, we identified 40 key knowledge products to review in depth, of which 31 are synthesized here (some studies had several products reporting similar results).

RESULTS

Question 1: What have we learned about how **governance changes related to land tenure rules and regulations** affect outcomes for women and contribute to gender equality?

A wide range of interventions and outcomes for women were considered in the 16 papers out of 26 included in this review that explicitly address this question in some manner. The rules, policies, regulations, and program-related interventions examined in each article are summarized below. These include those related to tenure, such as community-based legal aid programs, joint land titling, various women-targeted land tenure reform programs, land certification/registration programs, and collective/community-level forest governance and tenure reforms. Others primarily focused on livelihood-targeted interventions, including conditional cash transfers, women's collective microenterprises, community-based forest and wildlife enterprises, and social forestry. And others relate to organizational efforts such as Indigenous federations, women's rights and leadership forums, and gender-responsive land use planning approaches.

Evidence of various outcomes was also covered in these papers, and related to:

- **Inclusion/voice/participation**
  1. inclusion and participation in community-level forest governance or NRM governance bodies
- **Perceptions of tenure security**
  2. perceptions of the strength and security of women's land rights and rights to communal forests and forest resources, including for Indigenous women
- **Access to and control over natural resources and management practices, services, assets, and income**
  3. women's control over agricultural and forest-related resources/assets and/or income
  4. women's access to and control over solely and jointly held land
  5. women's adoption of new agricultural and/or natural resource management practices
- **Access to new opportunities/agency/empowerment**
  6. women acting in leadership roles in community forest governance
  7. changes in incidence and reporting of domestic violence
  8. creation of equitable employment opportunities
  9. increased roles of women in farmer-managed irrigation systems as a result of male migration

Evidence of longer-term impacts, such as impacts on women's income, food security, livelihoods, or poverty levels (at an individual level), are discussed in several cases but usually with little quantitative evidence. Table 1 summarizes the evidence found in these case studies from many countries regarding a range of governance interventions related to land tenure rules and regulations and how they affect various outcomes for women.[1]

The papers cited and discussed in the following sections all include some type of program, policy, or intervention and evidence of outcomes and impacts for women in particular and, in some cases, for gender relations more broadly. They are briefly summarized by the focus of the intervention(s) and cross-cutting lessons are then discussed.

---

1 A subset of 26 papers is included in Table 1: 16 looked at outcomes of interventions/governance/tenure reforms; 5 examined what gender analysis aimed at influencing governance and tenure reform shows; and 5 created frameworks/approaches and/or summarized available data and literature.

TABLE 1 COUNTRY CASE STUDY EVIDENCE BY TYPE OF INTERVENTIONS AND OUTCOMES

Outcome	Tenure-related interventions				Livelihoods-related interventions				Organizational-related interventions		
Changes in:	CB legal aid	Joint land titling	Land certif/ registration (women's title)	CB forest governance & tenure reforms	Conditional cash transfers	Women's microenterprises	CB forest & wildlife enterprises	Social forestry	Indigenous Federations	Women's rights & leadership fora	Gender-responsive land use plans, adaptive collabor. mgment
Inclusion and participation in community-level forest governance or NRM governance bodies				+				+	+		+
Perceptions of the strength and security of women's (incl. Indigenous) land rights & rights to communal forest resources	+		+	−				+	−	+	+
Women's leadership in community forest governance						−				+	+
Women's control over agricultural and forest-related resources, services, assets, income	+						+				
Women's access to and control over solely and jointly held land			+ −								
Incidence and reporting of domestic violence			+								+
Women's adoption of new agricultural and/or NRM practices							+				
Equitable employment opportunities							+				
Women's empowerment		−	−			+	−				
Women's NTFP income						+		+			

Source: Author's review. For citations by intervention type, see Section 3.1.

Note: + signifies positive outcomes noted; − are negative outcomes. A negative outcome for women's empowerment includes lack of empowerment or disempowerment. CB = community-based; NRM = natural resource management: NTFP = nontimber forest products

Women's rights to communal forests and community forest enterprises

Larson and colleagues (2019a) explore outcomes of community-level forest land tenure reforms in Peru. They investigate how national laws and community norms affect tenure security of collective lands for Indigenous women. In other words, in these collective lands, do individuals have the same rights (or tenure security) to the community resources? They find that important gender gaps exist in forest use, management, and decision-making, as well as in perceptions regarding the fairness of rules, tenure security, and drivers of insecurity related to titling and formalization processes. Across the 22 native communities studied, the results show that men participate more than women in communal forest management activities, with significant differences in the proportions participating in rulemaking (43 percent of men compared to 22 percent of women) and in forest monitoring (28 percent of men compared to 14 percent of women). In titled communities, 58 percent of male respondents perceived rules of resource use and forest access as fair, compared to only 36 percent of women. In addition, the proportion of men who perceived their rights as strong and secure (86 percent) was larger than the proportion of women (76 percent) (Larson et al. 2019a).

Because these are collective lands, women's rights depend not only on national-level policy and how titling is implemented, but also on community-level governance, which includes both considerations regarding if and how women's rights are locally recognized and the extent of their participation and influence in collective decision-making bodies. Key findings are that *how* women's rights are recognized by community leaders and members (if at all) and *how actively* women participate in collective decision-making bodies matters. Addressing these gender gaps is critical for tenure security outcomes. If women are not being included in forest landscape decision-making processes and empowered by the community, they will not experience secure tenure rights to forests and their products. The authors conclude that the following capacity strengthening actions are needed: facilitating women's participation in land tenure formalization processes; gender awareness for mainstreaming women's perspectives; and gender training and reflection for government, Indigenous federations, and communities.

Bose and colleagues (2017) synthesize five research studies, published as a special issue of a journal, and identify key policy priorities related to women's rights to communal forests, agricultural land, and natural resources in rural and Indigenous territories in five Latin American countries. In Colombia, joint land titling is not enough to assure Indigenous and marginal women's empowerment. In Nicaragua, there is evidence that women who control or participate in sales of forest products are more likely to control the income from them, yet women's participation and leadership in community governance is lacking. In Mexico, women's access to forest and agricultural lands is limited and joint titling is needed, as it also allows women to participate in community assemblies, hold elected posts, and share their views regarding natural resources in management plans for collective lands. The mixed effects on poverty levels of a conditional cash transfer poverty program in Mexico are also discussed. In the Brazilian Amazon, women's collective microenterprises were found to empower rural women and enhance their income. Training as well as environmental and technical knowledge improved women's natural resource management practices, gave them greater access to family land and resources, and increased their decision-making power over family land use. The authors suggest that closing the gender gap in rural farm and forest rights and assets will require participatory governance monitoring (such as adaptive collaborative management approaches), women's access to enterprise development and cash transfer programs, and joint titling coupled with support to women's entrepreneurship and access to markets.

Stoian and colleagues (2017) explore the outcomes and impacts of community forest enterprise (CFE) institutional arrangements/governance in Guatemala. The socioeconomic performance of these enterprises/groups was evaluated using the five capital assets (natural, social, physical, human, financial) framework to assess gender differences in asset endowments linked to CFEs. Gender outcomes examined include women's influence on household decisions, women's control over resources/assets, and women's participation in forest management. Impacts examined at the household level included household forest income and forest enterprise profits, and whether additional income allowed CFE members to rise above the poverty level. The authors find significant benefits/assets gained from these enterprises but with considerable variation across groups and households. Positive outcomes for women include increased decision-making power and control over resources, although again this varied across CFEs and households. The results also provide evidence of positive forest conservation and livelihood outcomes.

Outcomes and impacts of governance reforms related to forest tenure carried out through social forestry schemes[2] in Indonesia are examined by Siscawati (2020). The outcomes reported for social forestry programs include no significant change in gender norms, a gradual increase in women's participation in decision-making at household and community levels, greater security in management of the land, fewer men migrating, and more sustainable forest management. Positive impacts of social forestry schemes were also found for women's cash income from nontimber forest products and household income.

Gynch and colleagues (2020) examine the social, economic, and ecological outcomes of community-owned forest and wildlife enterprises in four countries (Guatemala, Mexico, Nepal, and Namibia). They describe a three-stage process of investment in community-owned resources that is underway. The first stage, triggered by rights devolution to communities, involves development of community user groups and sustainable resource management plans subject to government approval. In the second stage, CFEs are formed. The third stage occurs when the CFEs begin to perform well and attract new private investment.

The authors find evidence that these CFEs are delivering environmental and social returns as well as economic profits (Gynch et al. 2020). In relation to gender, they show that these enterprises have created equitable employment opportunities for poorer segments of the population. Implications for policies and programs include: 1) public sector and donor investments are critical in the first two stages, facilitating the entry of private sector financing in the third stage; 2) enabling investments are needed initially to strengthen institutions supporting communities, while asset investments come later; and 3) governments cannot retain their prior levels of control if social innovations by CFEs are to flourish.

Women's land tenure rights and tenure (in)security

Recognizing the importance of women's land rights, many countries have undertaken reforms, but the outcomes are not straightforward and depend on the initial conditions and status of women. In a book chapter, Ghebru (2019a) looks in depth at the effects of land governance reform on women's land rights in four African countries (Nigeria, Ethiopia, Mozambique, and Malawi). In three of the countries, Ghebru finds that the strength of women's tenure rights varies depending on how the parcel was acquired. In Mozambique (a relatively land-abundant country compared to Nigeria and Malawi), women have proportionally greater land rights over parcels acquired via customary sources (family inheritance or gift). In Nigeria and Malawi (relatively land-scarce countries), women's land rights are less constrained on purchased or rented parcels.

This work highlights the challenges of implementing reforms to ensure gender equality in land governance, including access to services (Ghebru

2 Social forestry schemes in the Indonesian context distinguish among those in community-managed forests, community-based plantation forests, and customary forest (Siscawati 2020).

2019a). Women face many constraints with respect to formal registration of customary land rights, including legal illiteracy (ignorance of land laws), social resistance to reforms, lack of political will, and high costs for land registration. Examples of innovative pro-women land tenure reform interventions in different countries are described, and include recognition of women's customary land rights, legal protection of women's individual use/occupancy rights, community land demarcation and collective titles, promotion of women's participation in decentralized land administration bodies, land registration/documentation, and different forms of decentralized dispute resolution mechanisms. The study finds that the main factors weakening women's land tenure rights include population pressure, commodification of land, and commercialization of agriculture (all increasing the value of land). Women's largely subsidiary and undocumented land rights under customary systems result in weak or insecure tenure levels. The limited incorporation of gender considerations in legal frameworks leads to poor implementation that does not contribute to gender equity.

Ghebru's chapter also summarizes the status of land governance reforms in 10 African countries (Ethiopia, Ghana, Madagascar, Malawi, Mozambique, Nigeria, Rwanda, Tanzania, Uganda, and Zambia), focusing on accessibility and sustainability of land rights protection interventions. The results show that, despite a range of gender-intentional reforms, little progress has been made toward achieving gender parity in land reform programs in any of the 10 countries. In terms of research gaps, the chapter suggests that analyses are needed of how the conjunction of gender with age, ethnicity, religion, and other factors affects both individual and joint land ownership, influences access to government services, affects empowerment, and is linked to domestic violence.

Land certification/registration

Ghebru (2019b) looks at changes in land rentals, incidence of land conflicts, and landholder incomes resulting from a gender-sensitive land certification program in Ethiopia. This program registers rights to individuals or households. The study examines gender outcomes that include women's access to and control over land, finding that women had greater access to and control over both solely and jointly held land in areas with the program compared to similar areas without the program.

Reporting on the same Ethiopian land certification program in its second stage (which consolidated more rights for those registered), Ghebru and Girmanchew (2020) use gender-disaggregated data from 1,984 households to examine the impact of the program on investments in land, women's land rights, land tenure security, and legal knowledge and awareness. They find a positive and statistically significant effect of the program on tenure security (both private and collective) for married women and a reduction in perceived private tenure risks for female household heads (single, divorced, and/or widowed) with land certificates. However, the findings also show that certification had a significant and negative impact on women and youth with respect to access to or control over land. This suggests that, by granting stronger exclusionary rights to the certificate holders (usually male and female heads of households), the program may not be protecting land rights of other women and youth.

In Nepal, Pradhan and colleagues (2019) examine women's empowerment, and how changes in property rights (such as through land registration efforts) may or may not influence it in many different contexts. Their study explores how Nepalese women's rights to personal and joint property, their life-cycle situation, and household social relationships affect their empowerment, particularly control over land and other assets. Results include findings that a woman's position within the household (for example, wife, mother-in-law, daughter-in-law) affects the strength and security of her property rights, and that this relationship is dynamic and changes over time. Women in different positions use different strategies to claim rights over personal and joint property. Property rights laws in Nepal do not account for such nuanced rights to assets shared within households. Land titles were found to be important for widows, however, as titles provided them some assurance that their family would continue to care for them as they aged. For others, making strong claims over personal assets can weaken their claims on joint property (land, house, etc.), which depend on being seen as a "good wife." Land registration efforts in the study area were found to have had a limited impact on women's empowerment, given the social risks associated with women registering land in their own names. The authors recommend that programs that transfer assets to women (such as land or livestock) need to take into account how women's positions and strategies with regards to land and other assets change over their life cycle.

Gender-responsive participatory village land use planning approaches

Two papers look at outcomes from a Tanzanian village land use planning approach called Women's Rights and Leadership Forums, which aims to increase the security of pastoral women's rights to land and natural resources (Kisambu et al. 2017; Dungumaro and Amos 2019). Training in women's rights to land, entrepreneurship, and leadership skills were provided, along with awareness training for male customary leaders on challenges faced by women and how to address them. Evidence demonstrates positive outcomes that include changes in perceptions and attitudes on the part of women and men (including recognition of women's rights in communities); issuance of individual land titles; a decrease in domestic violence; an increase in reporting of domestic violence; and an increase in the number of women leaders in the villages.

Other types of programs targeting land rights of women and indigenous communities

Billings and colleagues (2014) examine how community-based legal aid programs impact women's land rights, using a case study of a proposed legal aid bill in Tanzania. It concludes that these programs have a role to play in increasing access to services in rural areas, but that there is as yet no evidence on the impact on women's claims to land or more secure women's land rights.

Two studies (Evans et al. 2019; Evans et al. 2020) look at a community research and engagement model called adaptive collaborative management and the lessons learned from implementing it with Indigenous communities in Nicaragua. This project sought to enhance women's participation in community forest decision-making and governance. Through its "bottom-up" facilitated approach, the communities chose to co-develop tools for monitoring governance and leadership. The process was important for understanding criteria for good leadership and monitoring activities supported women's participation, but this alone was insufficient to address gendered tensions in communities (Evans et al. 2019). The authors emphasize the importance of social learning and reflection, especially with regard to gender relations, including among the research teams themselves (Evans et al. 2020).

Dupuits and Cronkleton (2020) examine governance changes in Peru associated with REDD+ that include Indigenous federations and the effect of these changes on outcomes including Indigenous tenure security and local participation in climate mitigation programs. The study also looks at how a government payment scheme (National Forest Conservation Program, PNCB) has impacted sustainable forest management and conservation efforts. It asks if tenure security is a major co-benefit of these REDD+ programs, and finds that, despite enhanced Indigenous Peoples' participation in these efforts, Indigenous tenure rights remain insecure in the absence of broader national programs for land tenure reform.

Summary of lessons learned: How changes in governance related to land tenure affect gender outcomes

While the interventions examined in these studies varied in scope, there are some cross-cutting lessons and implications for tenure policies and institutional arrangements that enhance gender equality.

First, women's and joint titling/certification must be complemented by community-level actions addressing gender tensions. National policies and titling/certification giving women and Indigenous Peoples rights to forests and land are insufficient on their own. Actions aimed at ensuring women are included in forest landscape decision-making processes and empowered within their communities are needed for them to experience secure tenure rights to forests and their products, as evidenced in Peru and Colombia. The need for a combination of interventions, beyond women's land certification, is also demonstrated in negative outcomes observed for women and youth in Ethiopia and Nepal. The adaptive collaborative management approach that supported women's participation in Nicaragua was also found to be insufficient on its own to address gendered tensions in communities. In Tanzania, combining interventions aimed at men (such as gender awareness training) with leadership and entrepreneurship trainings for women was found to lead to positive gender outcomes

Second, how the reforms are implemented and managed, and if and how gender is considered and addressed, affects whether gender outcomes are realized. Where women were not involved in titling and communities still largely excluded women, as in the case of Peru, gender-related outcomes were not positive. The experience in Tanzania, where gender awareness trainings for men were included, was more positive.

It also clear that many of these governance approaches have achieved outcomes related to enhanced and more equitable participation of women but have not yet resulted in increased women's benefits or empowerment. In applying their "reach, benefit, empower" framework, Johnson and colleagues (2018) also found many agricultural projects seeking to empower women fell short, succeeding only in achieving enhanced participation – a necessary first step, perhaps, but insufficient by itself.

Question 2: What do **gender analyses** aimed at informing tenure or governance reforms show?

Five of the papers reviewed undertook some type of gender analysis in order to inform gender-responsive tenure or governance-related reform efforts. They include studies from Uganda, Peru, Indonesia, Ghana, Nigeria, and Mozambique.

Monterroso and colleagues (2019) address the underlying conditions necessary for governance or policy reforms in collective forest tenure to result in better outcomes for women and other marginalized groups. The multicountry comparative study looks at the impacts of forest governance reforms on gender equality, forest conditions, livelihoods, and tenure security. The authors use the Doss and Meinzen-Dick (2018) framework to analyze the context of existing threats and opportunities as catalysts of change in collective forest tenure policies. They conduct a gender analysis to better understand the conditions that enhance women's ability to participate and benefit from collective forest tenure reforms in Uganda, Peru, and Indonesia, finding that beliefs and social norms within communities are a source of tenure insecurity for women. The study compares outcomes of reforms for women versus men. Outcomes examined include perceptions of tenure security, increased participation, and adoption of improved management practices. Impacts include improved income and livelihoods. Key findings of the study are that, following the reforms, women are participating in meetings around forest management, but often with no decision-making power (that is, there is a positive impact on the inclusion outcome but negative impact for empowerment). The governance reforms have also increased women's adoption of new management practices. The authors conclude that improving extension services to meet the needs and activities of women could lead to better livelihood and forest condition outcomes and impacts.

Lambrecht (2016) explores factors influencing women's access to agricultural land in Ghana and the implications for policy and governance reforms. She finds that social norms, perceptions, and formal and informal rules shaping access to land for female and male farmers are important at the household, community, market, and national levels. The results show that, "norms related to household and family organization and on men's and women's responsibilities and capabilities play a key role in gendered allocation of resources. However, these norms and perceptions are dynamic and evolve jointly with the development of markets and changes in values of inputs such as labor and land" (Lambrecht 2016: 188). Because of this, Lambrecht suggests that policies and laws addressing inheritance and gender-equitable distribution of property will be critical for strengthening women's property rights, but that strengthening women's property rights will remain challenging in areas where patriarchal customary rights and social norms and rules persist.

Also in Ghana, Ghebru and Lambrecht (2017) undertook an econometric analysis to explore the factors associated with farmers' perceptions of tenure security at the household and individual plot levels. They find that gender, social status, and other social and economic factors are all drivers of tenure insecurity for individually held or household-owned land. Individual farmers (women and men) feel more secure on purchased or inherited land than they do on land allocated by traditional authorities. In communities with active land and agricultural markets, households feel less secure in their land rights. Those who feel the least secure include migrant households and women in polygamous households, while farmers with political connections feel the most secure.

Ghebru and Girmachew (2017) investigate determinants of land tenure insecurity in Nigeria. Gender is included by looking at factors influencing perceptions of tenure security for female-headed households and for female-managed plots. The results show that perceived risks of private land disputes are higher for female- than male-headed households. Tenure security is perceived to be relatively higher on female-managed plots of female-headed households than on female-managed plots of male-headed households. They conclude that systematic land titling and registration should be implemented at the parcel level because household-based land titling does not account for intrahousehold dynamics.

Gender-disaggregated data are also analyzed to improve understandings of inter- and intra-household differences in causes of tenure insecurity in Mozambique (Ghebru and Girmachew 2019). This study uses two proxy measures for land tenure insecurity – perceived risk of land loss (collective tenure risk) and perceived risk of a private land dispute (individual tenure risk). The evidence suggests that fear of land loss is perceived as the greatest threat to women's tenure security. For men, individual risks such as ownership and inheritance (land disputes) are considered larger threats to their tenure security. However, fear of individual tenure threats is higher for female spouses compared to males within the same household. The results also show that female spouses who have no control over family land are more likely to feel insecure. Women who are more aware of legal issues around land rights report greater tenure security. The authors also find that tenure insecurity for female spouses is associated with the emergence of land markets, but for female household heads, it is relative land scarcity in a given community that drives tenure insecurity. They conclude that since gender-related differences exist in perceived tenure risks, programs aimed at enhancing tenure security need to be more targeted and include measures to address these different concerns.

Summary of lessons learned and policy implications: What gender analyses aimed at informing tenure or governance reforms show

These insights into common causes of women's and men's perceived (in)security of land tenure – including a fear of losing one's land, risks related to land disputes, and restrictive social norms and traditional tenure regimes – point to several direct implications for policies and programs, such as the need for investment in actions to raise men and women's awareness of women's land rights. Together, these analyses, drawn from diverse countries, reinforce the message that gender differences in perceived land tenure risks are widespread. Thus, informed governance choices will likely be much more effective in achieving outcomes that both reach more women (more equitable participation in governance) and foster adoption of improved management practices by both women and men, thus supporting longer-run impacts such as higher incomes and improved livelihoods.

Another key lesson this research highlights is that even when women are being reached and are participating in community-level forest and natural resource decision-making bodies, they often remain without decision-making power, which limits the benefits they reap and impedes empowerment and gender equality outcomes. Gender-responsive rural advisory services, including advice on property rights, are needed along with policies and laws addressing gender-equitable inheritance and distribution of property to strengthen women's property rights and promote other desired gender outcomes.

These studies also reveal commonalities in terms of the role of state law versus social norms, pointing to the need for joint approaches that address the governance obstacles to women's tenure rights in tandem. Such approaches will differ in different contexts and circumstances, reinforcing the need for gender analyses to inform local and national interventions focused on land and natural resource tenure.

Question 3: What **approaches, methods, tools, frameworks, and data** are available for addressing the gender, landscape governance, and tenure nexus?

Nine of the knowledge products reviewed here developed new frameworks, approaches, and tools or synthesized available data relevant to these issues. As mentioned in the introduction, Doss and Meinzen-Dick (2018, 2020) developed a conceptual framework to examine the factors influencing women's land rights and tenure security. They argue that a clearer definition of women's tenure security is needed in order to understand factors influencing tenure security in different contexts, and therefore, what governance or other changes are needed to strengthen it. The framework covers key dimensions of the context (regional, national, local) and the dynamic factors that may affect (strengthen, undermine) women's tenure security. The framework is aimed at practitioners and researchers, with the goal of establishing common indicators and supporting collection of more consistent data to allow for comparative analyses. This framework is starting to be used to produce the kind of evidence needed by policymakers to formulate evidence-based programs and sustainable development-oriented solutions (see, for example, Monterosso et al. 2019).

Meinzen-Dick and colleagues (2017) explore the connections between women's land rights and poverty through a literature review. They focus on understanding how strengthening women's land rights affects poverty reduction efforts and the well-being of women and their households in rural areas. The paper identifies the intermediate outcomes along the different pathways to poverty reduction through interventions aimed at strengthening women's land security and examines the empirical evidence for each pathway. They find strong evidence across the literature that strengthened land rights improve outcomes related to bargaining power and decisions regarding consumption, human capital investment, and intergenerational transfers. There is agreement but weaker evidence that women's land rights have a significant influence on (or are influenced by) outcomes such as uptake of improved natural resource management practices, strengthened government services and institutions, women's empowerment and reduced domestic violence, enhanced resilience and HIV risk, or consumption and food security. There is limited evidence on associations between women's land rights and credit, technology adoption, and agricultural productivity.

The authors also find no papers that directly investigate the link between women's land rights and poverty, and they call for more rigorous evaluations of reforms that strengthen these rights. They suggest that "many of the evidence gaps arise from a failure to account for the complexity of land rights regimes, the measurement of land rights at the household level, the lack of attention paid to gender roles, and the lack of studies from countries outside Africa" (Meinzen-Dick et al. 2017: v). The authors conclude that ongoing and increased efforts to design and implement programs and policies to strengthen women's land rights need to be supported and different approaches tested with appropriate counterfactuals.

Kieran and colleagues (2015) review data on men's and women's control of land in four Asian countries (Bangladesh, Tajikistan, Timor-Leste, and Viet Nam) to assess knowledge gaps regarding the land rights of women and men. Five measures or indicators are used: incidence of land ownership, distribution of land ownership, distribution of plots owned, mean plot size, and distribution of land area. The available data suggest there are large proportional differences in women's and men's land ownership in all four countries. However, there is limited information on joint and individual ownership, thus more effort is needed in data collection on intra-household tenure rights.

McLain and colleagues (2018) develop an approach to support integration of gender, tenure, and governance considerations into forest landscape restoration efforts. They suggest the use of a rights-enhanced approach to forest landscape restoration to support stronger and more secure tenure rights for the landholders involved – meaning that these approaches must implement actions that provide women and other marginalized groups with rights to land and trees. The authors' recommended tenure diagnostic approach is intended to complement existing planning approaches forest restoration, such as the Restoration Opportunities Assessment Methodology (ROAM), a gender-integrated decision-support tool that is now widely used in national and subnational planning (IUCN and WRI 2014).

Jhaveri (2020) provides practical guidance to practitioners aiming to support gender-responsive community-based forest tenure reforms, drawing from a wide range of sources. She includes conceptual ideas, operational direction, examples of good practices and case study insights, research findings, and information on available resources. The guide describes different pathways to achieving gender equality in forest tenure and promotes a three-stage approach consisting of: 1) analysis of gender challenges and opportunities; 2) inclusive and forward-looking strategic planning; and 3) creating and implementing actions for realizing pathways that close identified gender gaps.

The Center for International Forestry Research (CIFOR) has developed a toolbox with several approaches that support the inclusion of marginalized groups in multistakeholder forums and processes around landscape governance and management.[3] These tools were co-developed with diverse actors from government, Indigenous, and local communities; women's organizations; and NGOs. The main tool, called How Are We Doing?, involves participatory monitoring and reflective learning with diverse actors in the landscape who affect or are affected by the issue being examined. Strategies for active participation of those often excluded from decision-making processes, such as women, are encouraged (Sarmiento Barletti et al. 2020a).

A second tool, ¿Cómo vamos?, was developed collaboratively with Peru's National Organization of Andean and Amazonian Indigenous Women (ONAMIAP) (Sarmiento Barletti et al. 2020b). This tool enables Indigenous women and their allies to reflect on how their communities are governed and to collectively find solutions for their own inclusion in those processes. A targeted set of indicators was developed and tested with Indigenous women in the Peruvian Amazon and Andes.

A third set of tools, called Getting It Right, comes in the form of a guide that explains how to operationalize the inclusion of women, Indigenous Peoples, and other underrepresented groups in

3 https://www.cifor.org/toolboxes/tools-for-managing-landscapes-inclusively/

multistakeholder forums (Evans et al. 2021). It takes a rights-based approach to show multistakeholder groups how to build practical, flexible monitoring approaches that promote inclusion.

Summary of lessons learned: Approaches and frameworks

In most countries, large gaps remain in high-quality information on the gender, governance, and tenure nexus. There is no single framework to guide researchers and practitioners tackling challenges in this nexus as they work to support sustainability and poverty reduction. However, several conceptual frameworks developed or adapted by PIM5 researchers provide ways to measure progress toward improved tenure security and governance with a gender angle. Doss and Meinzen-Dick (2020) focus on identifying where governance reforms are needed in order to strengthen women's rights; Stoian and colleagues (2017) assess gender differences in assets related to community forest enterprises; and McLain and colleagues (2018) offer a more equitable, rights-based approach to forest landscape restoration. Other approaches and tools support processes for fostering progress, such as Jhaveri (2020) on gender and tenure and CIFOR's social inclusion toolbox for multistakeholder processes (Sarmiento Barletti et al. 2020a, 2020b; Evans et al. 2021).

These approaches demonstrate the complexity of women's tenure (in)security for land and natural resources and that context matters, as does an understanding of dynamic factors that may strengthen or undermine security. These contextual factors also come into play in informing what governance processes, policies, strategies, programs, and so on are best suited to achieving gender outcomes that go beyond reach or participation and lead to empowerment.

The tools are aimed at increasing the availability of consistent, gender-disaggregated data and common indicators for comparative analysis and for monitoring progress toward multiple outcomes. More widespread use of these tools, along with institutionalizing collection of gender-disaggregated data and common indicators, will be critical for identifying specific, evidence-based policies and practices that lead to gender equality in diverse landscapes.

CONCLUSIONS

This synthesis has focused on a relatively small subset of the research work supported by PIM5 in the broad areas of gender, landscape governance, and tenure, focusing to the extent possible on the overlap of the three. This body of work has added to the knowledge base related to this nexus of issues regarding gender-related outcomes of interventions, governance, and tenure reforms (16 out of 26 papers reviewed), and what gender analyses aimed at informing governance and tenure reforms reveal in different country contexts (5 papers). Overall, the empirical studies show the complexity of the processes involved in securing women's land rights or gender-equitable governance, and have contributed to understanding in particular cases. The frameworks developed help conceptualize key issues and research questions. There are also several new approaches, tools, and methods now available that will improve practitioners' and policymakers' ability to design evidence-based policies and interventions that enhance gender equality (9 papers).

The work provides evidence that a wide range of governance and policy reforms aimed at improving women's and men's landscape management practices and strengthening their land and natural resource tenure security are contributing to multiple positive outcomes and impacts in many countries. A focus on natural resource governance, institutions, and rights is not just beneficial for women, but also for diverse marginalized groups, as the collaborative approaches and tools developed with Indigenous Peoples demonstrate (Sarmiento Barletti et al. 2020a, 2020b; Evans et al. 2021).

Several of these studies make the case that better governance approaches (such as adaptive collaborative management approaches at the local level and national laws that encompass equity and social inclusion), attention to tenure rights (such as joint titling), and a focus on livelihoods (such as women's access to enterprise development and markets, support to women's entrepreneurship, and cash transfer programs) are all needed for the advancement of gender equality (see, for example, Bose et al. 2017; Gynch et al. 2020).

The research focusing on gender analyses – looking at what factors drive women's and men's tenure (in)security and at the different sources of tenure insecurity for men and women – demonstrates that solutions aimed at strengthening tenure security may differ for men and women. The analyses provide evidence that understanding gender norms and local contexts is critical for informing land and natural resource tenure-focused interventions. They also point to commonalities in the key, but insufficient role of national laws that promote equity and social inclusion; these laws are insufficient because social norms and customs at the local and community level continue to have a major influence. This indicates the need for joint approaches that also address these norms and rules. Approaches for doing so will differ by context and circumstances, reinforcing the need for gender analyses to inform both local and national interventions. Encouragingly, where national laws (such as forest laws) and subnational laws both adopt equity and social inclusion as key principles, more progress toward gender equality goals is seen, but many implementation challenges remain (see Peru in Larson et al. 2019b; India, DRC, and Nepal in Jhaveri 2020).

The importance of paying attention to how governance rule changes and policy reforms are being implemented is another cross-cutting lesson (Larson et al. 2019; Jhaveri 2020), as it can shed light on opportunities to address gender inequalities. For example, formalizing community rights to forests in Peru has allowed women to be recognized in communal bylaws by establishing mechanisms for their participation in collective decisions around forest resources (Larson et al. 2019b).

This review points to several areas for future research, including the implications of different governance, gender, and tenure-related interventions on food security (see, for example, Larson et al. 2019b) and investments by different actors (private and public sector) in landscape restoration practices (McLain et al. 2018). Most relevantly, there is still much research to be done on the complex nexus of gender, tenure security, and landscape governance, with multiple pathways and policy actions that could potentially contribute to more sustainable natural resource management and more equitable poverty reduction. Future work may benefit from the development of more comprehensive theories of change that make these outcome pathways clearer and the influence of interventions on specific outcomes and impacts more obvious. This will require more rigorous impact evaluations, for example, on the impacts of tenure security-enhancing interventions. However, analyses that could provide such evidence are difficult as the interventions are seldom standalone and the pathways complex – that is, they are linked to other actions and factors that together contribute to overall outcomes – and gender equality is particularly challenging to measure.

AUTHOR

Patricia Kristjanson, PhD, is an agricultural economist whose expertise includes poverty, livelihood and gender analyses, agricultural landscape and food systems analyses, and innovative research approaches for linking knowledge with action. Now an independent consultant, she has over 30 years of experience leading and managing interdisciplinary agriculture research for development teams from international and national agricultural research centers, in collaboration with universities, development agencies, and private and public sectors across Africa, Asia, and Latin America.

REFERENCES

Agarwal, B. 1994. *A Field of One's Own: Gender and Land Rights in South Asia*. Cambridge: Cambridge University Press.

Agarwal, B. 1997. "'Bargaining' and Gender Relations: Within and Beyond the Household." *Feminist Economics* 3 (1): 1-51.

Billings, L., R.S. Meinzen-Dick, and V. Mueller. 2014. *Implications of Community-Based Legal Aid Regulation on Women's Land Rights*. IFPRI Research Brief 20. Washington, DC: International Food Policy Research Institute (IFPRI).

Bose, P., A.M. Larson, S. Lastarria-Cornhiel, C. Radel, M. Schmink, B. Schmook, and V. Vázquez-García. 2017. "Women's Rights to Land and Communal Forest Tenure: A Way Forward for Research and Policy Agenda in Latin America." *Women's Studies International Forum* 65: 53-59.

Buck, L.E., S.J. Scherr, B. Chami, M. Goldman, T. Lawrence, J. Mecham, E. Nevers, and R. Thomas. 2019. *Exploring Property Rights and Tenure in Integrated Landscape Management*. Washington, DC: EcoAgriculture Partners on behalf of the Landscapes for People, Food and Nature Initiative.

Colfer, C.J.P., B. Sijapati Basnett, and M. Elias, eds. 2016. *Gender and Forests. Climate Change, Tenure, Value Chains and Emerging Issues*. London: Earthscan.

De Graaf, M., L. Buck, S. Shames, and R. Zagt. 2017. *Assessing Landscape Governance: A Participatory Approach–Manual*. Wageningen, Netherlands: Tropenbos International; Washington, DC: EcoAgriculture Partners.

Doss, C.R., and R.S. Meinzen-Dick. 2018. *Women's Land Tenure Security: A Conceptual Framework*. Seattle, WA: Research Consortium.

Doss, C.R., and R.S. Meinzen-Dick. 2020. "Land Tenure Security for Women: A Conceptual Framework." *Land Use Policy* 99: 105080.

Dungumaro, J., and M. Amos. 2019. *Pastoral Women's Rights and Leadership Forums, Tanzania: Experience, Impact and Lessons Learned*. Rangelands Research Report 3. Nairobi: International Livestock Research Institute (ILRI).

Dupuits, E., and P. Cronkleton. 2020. "Indigenous Tenure Security and Local Participation in Climate Mitigation Programs: Exploring the Institutional Gaps of REDD+ Implementation in the Peruvian Amazon." *Environmental Policy and Governance* 30 (4): 209-220.

Evans, K., S. Flores, and A.M. Larson. 2019. "Participatory Monitoring in Forest Communities to Improve Governance, Accountability and Women's Participation." *Small-Scale Forestry* 18: 165-187.

Evans, K., A.M. Larson, and S. Flores. 2020. "Learning to Learn in Tropical Forests: Training Field Teams in Adaptive Collaborative Management, Monitoring and Gender." *International Forestry Review* 22 (2): 189-198.

Evans, K., I. Monterroso, D.B. Ombogoh, N. Liswanti, A. Tamara, H. Mariño, J.P. Sarmiento, and A.M. Larson. 2021. *Getting It Right, A Guide to Improve Inclusion in Multi-stakeholder Forums*. Bogor, Indonesia: CIFOR.

FAO (Food and Agricultural Organization of the United Nations). 2011. *The State of Food and Agriculture. Women in Agriculture: Closing the Gender Gap for Development*. Rome.

FAO. 2019. "Assessing the Governance of Tenure for Improving Forests and Livelihoods: A Tool to Support the Implementation of the Voluntary Guidelines on the Responsible Governance of Tenure." Forestry Working Paper, no. 13. Rome.

Feldstein, H.S., and S.V. Poats, eds. 1990. *Working Together: Gender Analysis in Agriculture*. West Hartford, CT: Kumarian Press.

Fortmann, L., and D. Rocheleau. 1989. *Why Agroforestry Needs Women: Four Myths and a Case Study, in Women's Role in Forest Resource Management: A Reader*. Bangkok: FAO Regional Office for Asia and the Pacific.

Ghebru, H. 2019a. "Women's Land Rights in Africa." In *2019 RESAKSS Annual Trends and Outlook Report: Gender Equality in Rural Africa: From Commitments to Outcomes*, eds. A.R. Quisumbing, R.S. Meinzen-Dick, and J. Njuki, Chapter 4, 44-56. Washington, DC: IFPRI.

Ghebru, H. 2019b. "The Impacts of Second-Level Land Certification (SLLC) in Ethiopia: Empirical Evidence Using Panel Data." Presented at the *World Bank's 20th Annual Land and Poverty Conference* in Washington, DC, March 25-29.

Ghebru, H., and F. Girmachew. 2017. "Scrutinizing the Status Quo: Rural Transformation and Land Tenure Security in Nigeria." NSSP Working Paper 43. International Food Policy Research Institute, Washington, DC.

Ghebru, H., and F. Girmachew. 2019. "Perceived Tenure (In)Security in the Era of Rural Transformation: Gender-Disaggregated Analysis from Mozambique." IFPRI Discussion Paper 1799. International Food Policy Research Institute, Washington, DC.

Ghebru, H., and F. Girmachew. 2020. "Direct and Spillover Effects of Land Rights Formalization in Africa: A Case Study of the Second-Level Land Certification (SLLC) in Ethiopia." IFPRI Discussion Paper 1968. IFPRI, Washington, DC.

Ghebru, H., and I. Lambrecht. 2017. "Drivers of Perceived Land Tenure (In)Security: Empirical Evidence from Ghana." *Land Use Policy* 66: 293-303.

Giovarelli, R., and E. Scalise. 2019. *Women's Land Tenure Framework for Analysis: Land Rights*. Friday Harbor, WA: Resource Equity.

Gynch, S., S. Lawry, R. McLain, I. Monterroso, and A. Adhikary. 2020. "Is Community Tenure Facilitating Investment in the Commons for Inclusive and Sustainable Development?" *Forest Policy and Economics* 111: 102088.

IUCN (International Union for Conservation of Nature) and WRI (World Resources Institute). 2014. "A Guide to the Restoration Opportunities Assessment Methodology (ROAM): Assessing Forest Landscape Restoration Opportunities at the National or Sub-national Level." Working Paper (Road-test edition). IUCN, Gland, Switzerland; WRI, Washington, DC.

Jhaveri, N.J. 2020. *Forest Tenure Pathways to Gender Equality: A Practitioner's Guide*. Bogor, Indonesia: Center for International Forestry Research (CIFOR).

Joekes, S., M. Leach, and C. Green, eds. 1995. "Gender Relations and Environmental Change." *IDS Bulletin* 26 (1): 1–100.

Johnson, N., M. Balagamwala, C. Pinkstaff, S. Theis, R. Meinzen-Dick, and A. Quisumbing. 2018. "How Do Agricultural Development Projects Empower Women? Linking Strategies with Expected Outcomes." *Agri-Gender* 3 (2): 1–10.

Kieran, C., K. Sproule, C. Doss, A.R. Quisumbing, and S.M. Kim. 2015. "Examining Gender Inequalities in Land Rights Indicators in Asia." *Agricultural Economics* 46 (1): 119–138.

Kisambu, N., F. Flintan, E. Daley, and S. Pallas. 2017. "Gender and the Commons: Pastoral Women's Land Rights and Village Land Use Planning in Tanzania: Experiences from the Sustainable Rangeland Management Project." Paper presented at the Conference of the International Association for the Study of the Commons, Utrecht, the Netherlands, July 10–14, 2017.

Lambrecht, I. 2016. "'As a Husband I Will Love, Lead, and Provide.' Gendered Access to Land in Ghana." *World Development* 88: 188–200.

Larson, A.M., I. Monterroso, and P. Canturias. 2019a. *Gender and Formalization of Native Communities in the Peruvian Amazon*. CIFOR Infobrief no. 238. Bogor, Indonesia: CIFOR.

Larson, A.M., I. Monterroso, N. Liswanti, T. Herawati, A. Banana, P. Cantuarias, K. Rivera, and E. Mwangi. 2019b. *Models for Formalizing Customary and Community Forest Lands: The Need to Integrate Livelihoods into Rights and Forest Conservation Goals*. CIFOR Infobrief no. 253. Bogor, Indonesia: CIFOR.

McLain, R., S. Lawry, M.R. Guariguata, and J. Reed. 2018. "Toward a Tenure-Responsive Approach to Forest Landscape Restoration: A Proposed Tenure Diagnostic for Assessing Restoration Opportunities." *Land Use Policy* 104: 103748.

McLain, R., S. Lawry, M.R. Guariguata, and J. Reed. 2019. *Integrating Tenure and Governance into Assessments of Forest Landscape Restoration Assessments*. CIFOR InfoBrief no. 241. Bogor, Indonesia: CIFOR.

Meinzen-Dick, R., L.R. Brown, H. Sims Feldstein, and A.R. Quisumbing. 1997. "Gender, Property Rights, and Natural Resources." *World Development* 25 (8): 1303–1316.

Meinzen-Dick, R.S., A.R. Quisumbing, C.R. Doss, and S. Theis. 2017. "Women's Land Rights as a Pathway to Poverty Reduction: A Framework and Review of Available Evidence." IFPRI Discussion Paper 1663. IFPRI, Washington, DC.

Monterroso, I., A. Larson, E. Mwangi, N. Liswanti, and Z. Cruz-Burga. 2019. *Mobilizing Change for Women within Collective Tenure Regimes*. Bogor, Indonesia: CIFOR.

Pradhan, R., R. Meinzen-Dick, and S. Theis. 2019. "Property Rights, Intersectionality, and Women's Empowerment in Nepal." *Journal of Rural Studies* 70: 26–35.

Quisumbing, A.R., and N. Kumar. 2014. "Land Rights Knowledge and Conservation in Rural Ethiopia Mind the Gender Gap." IFPRI Discussion Paper 1386. IFPRI, Washington, DC.

Sarmiento Barletti, J.P., A.M. Larson, N. Cisneros, N. Heise, N. Liswanti, H. Mariño, and A. Tamara. 2020a. *How Are We Doing? A Tool to Reflect on the Process, Progress and Priorities of Your Multi-stakeholder Forum*. Bogor, Indonesia: CIFOR.

Sarmiento Barletti, J.P., H.S. Mariño Saavedra, M. Bustamante de Almenara, N.B. Aguilar Alvarado, S.M. Canales Poma, and A.M. Larson. 2020b. *¿Cómo vamos? Una herramienta para reflexionar sobre la participación de las mujeres indígenas u originarias en la gestión y gobernanza de sus territories*. Lima: Organización Nacional de Mujeres Indígenas Andinas y Amazónicas del Perú; Bogor, Indonesia: CIFOR.

de Siqueira, L.P., A.M. Tedesco, P. Meli, A. Diederichsen, and P. Brancalion. 2021. "Gender Inclusion in Ecological Restoration." *Restoration Ecology* 29 (7): e13497.

Sijapati Basnett, B., M. Elias, M. Ihalainen, and A.M. Paez Valencia. 2017. *Gender Matters in Forest Landscape Restoration A Framework for Design and Evaluation*. CIFOR Brief 6685. Bogor, Indonesia: CIFOR.

Siscawati, M. 2020. "Gender and Forest Tenure Reform in Indonesia." CIFOR Working Paper No. 258. CIFOR, Bogor, Indonesia.

Stoian, D., M. Elias, and A. Rodas. 2017. "Land, Livelihoods and Gender: Dynamics of Community Forestry in Peten, Guatemala." In *Tropentag 2017–Future Agriculture: Social-ecological Transitions and Bio-cultural Shifts: Book of Abstracts of the International Conference Held in Bonn, Germany on September 20-22, 2017*, eds. E. Tielkes, 504.